31 – 1st Grade Math Puzzles

Practice various 1st Grade math
skills by putting the puzzles together.

I0426945

Addition, subtraction, missing addends, telling time,
commutative and associative properties of addition,
& more are included!

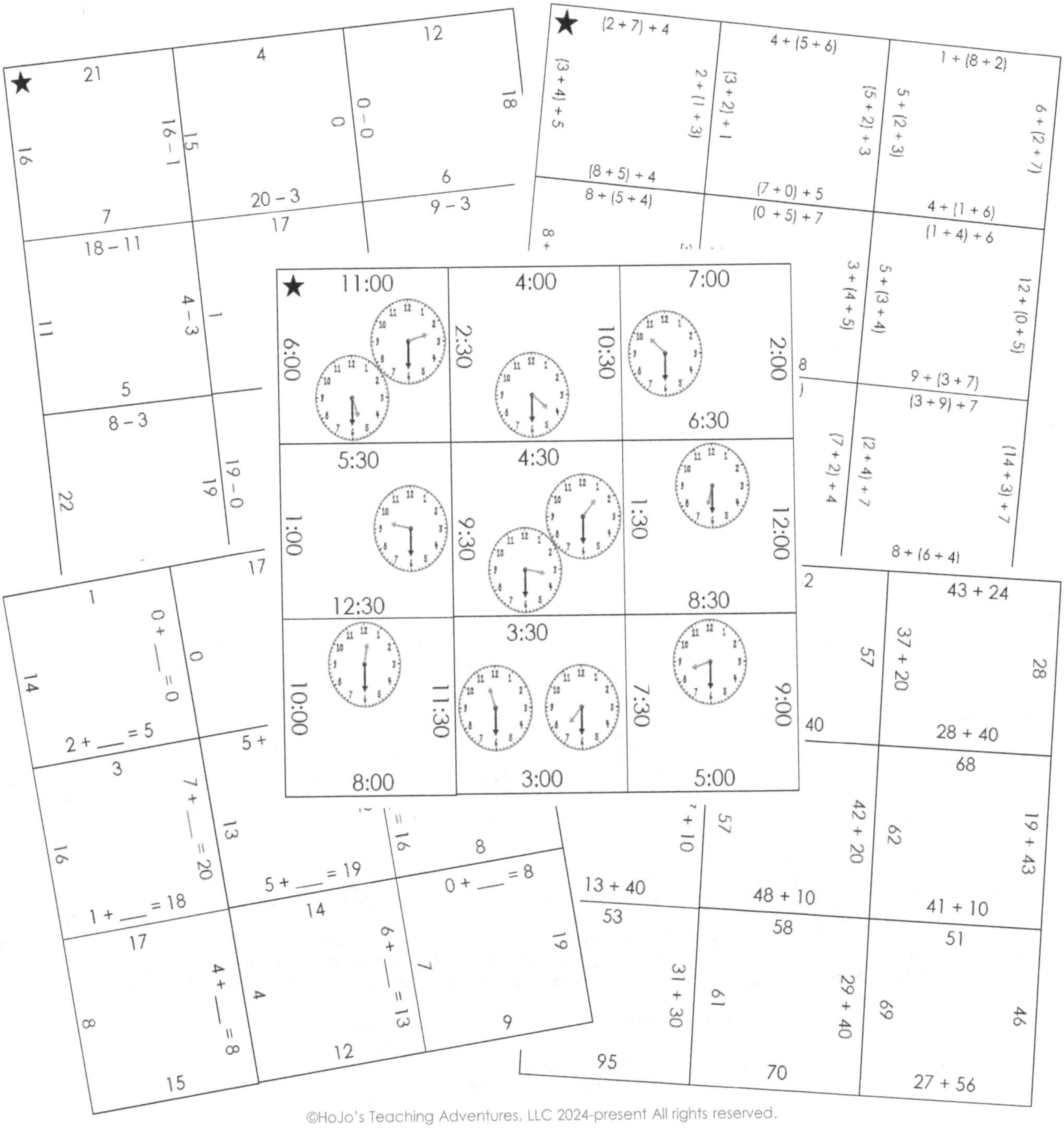

1

Table of Contents	Page
Matching Numbers to Number Words #1-12	7
Matching Numbers to Number Words #13-24	9
Matching Ten Words to Numbers 10-120	11
2D Shapes	13
3D Geometric Shapes	15
Adding within 10	17, 19
Subtracting within 10	21, 23
Adding two digit numbers + multiples of 10 – Adding by 10, 20, or 30	25
Adding two digit numbers + multiples of 10 – Adding by 10, 20, 30, or 40	27
Adding two digit numbers + multiples of 10 – Adding by 50, 60, or 70	29
Associative Property of Addition	31, 33, 35, 37
Missing Addends of Addition within 20	39, 41
Missing Addends of Subtraction within 20	43, 45
Addition 0-20	47, 49
Subtraction 0-20	51, 53
Telling Time to the Nearest Hour	55
Telling Time to the Nearest Half Hour	57
Commutative Property of Addition	59, 61, 63
Adding within 100 (Two Digit Number + One Digit Number)	65, 67
3x3 Puzzle Mat	69
Answer Key	71-76

Instructions for Use

These puzzles were created to help 1st Grade students practice various math skills in a **hands-on, engaging,** and **FUN** format!

Choose the puzzle you want, cut on the dotted line, cut out the nine puzzle pieces, and hand it to the child. The child will put the 9-piece puzzle back together.

If you want a bit more support for the child, also give them the puzzle mat so they can see the square shape and where the first puzzle piece with the star goes. The puzzle mat is <u>optional</u>, but it does offer more support for kids who are new to these puzzles or those who need a bit more help as they work through the math skills.

There are **31 different math puzzles** to choose from, so you're sure to have a variety of options to work on 1st Grade math all year long!

<u>Ideas for Extra Support:</u>
- Explain to the child that the puzzle piece with the star goes in the upper left-hand corner.
- Make sure they use the included puzzle mat! (The star piece will be easily identifiable this way.)
- Work through the puzzles <u>with</u> the child the first time.

<u>Extended Learning:</u>
- Save the puzzles and let students complete them more than one time.
- Have students create their own puzzle using a square shape with nine pieces. Cut and you'll have another puzzle to put together.

Answer keys for all 31 puzzles have been included at the end of this book.

I hope you enjoy these puzzles as much as I have enjoyed creating them! This teacher turned mom THANKS YOU for your interest in my work! Happy learning!!

★twenty-four 17 … 1 nine	15 one … 10 8	eighteen ten … 20 2
9 twenty-one … eleven five	eight 11 … four 12	two 4 … sixteen 7
5 14 … three twenty-three	twelve 3 … six 19	seven 6 … 22 thirteen

Cut here ✂

★ four 7 thirteen 21	12 13 23 seventeen	one twenty-three 9 15
twenty-one two 19 fourteen	17 nineteen twenty-two 24	fifteen 22 six 20
14 11 eighteen five	twenty-four 18 sixteen 8	twenty 16 10 three

Cut here ✂

★ six / thirteen / 30 / sixty	one / thirty / seventy / 40	eleven / 70 / eight / ten
60 / two / one hundred / 100 / 90	forty / one hundred twenty / 50	10 / 120 / five / one hundred ten / 110
ninety / seven / 20 / twelve	fifty / twenty / four	110 / eighty / 80 / three / nine

★ red	white	blue
pink · square	square · trapezoid	octagon · orange · heart
hexagon	parallelogram	heart · triangle · brown
yellow · parallelogram	trapezoid · rectangle	triangle · oval
pentagon	rectangle	oval · rhombus · gray
white · green	circle · purple	rhombus · black

★ pentagon parallelogram sphere	square triangular pyramid hemisphere	triangle pentagonal prism oval
trapezoid triangular prism rectangular prism	hemisphere cone rectangular prism	right pyramid rectangle
rectangular prism circle octagon	cube hexagon	right triangular prism cylinder diamond rhombus

★ 11 20 2 + 1 3 + 1	17 3 5 1	14 3 + 2 19 6
4 15 2 5 + 3	0 + 1 1 + 1 9 0 + 0	3 + 3 5 + 4 12 2 + 5
8 18 6 13	0 2 + 4 10 21	7 5 + 5 22 16

Cut here ✂

★ 20	14	18
11 ... 2 + 8 ... 2 + 2	10 ... 2 ... 3 + 4	0 + 2 ... 12 ... 7 + 2
4	7	9
16 ... 6 + 3 ... 1	9 ... 8 ... 6	4 + 4 ... 15 ... 3 + 7
1 + 0	4 + 2	10
21 ... 1 + 2 ... 13	3 ... 4 + 1 ... 17	5 ... 19 ... 0

★ 17 12 — 4−2 — 5−2	22 2 — 0 — 9	15 1−1 — 19 — 1
3 16 — 0 — 9−2	10−1 10−0 — 4 — 7−1	3−2 5−1 — 11 — 5−0
7 21 — 1 — 18	6 10−9 — 10 — 14	5 10−0 — 13 — 20

★ 22	15	13
16 · 6−4 · 8−2	2 · 4 · 6−5	10−6 · 19 · 7−4
<u>6</u>	1	3
12 · <u>9−6</u> · 0	3 · 5 · 7	8−3 · 14 · <u>8−6</u>
7−7	10−3	2
20 · <u>9−0</u> · 17	<u>9</u> · 10−2 · 18	8 · 21 · 11

Piece 1: top: 21 + 52 | left: 58 | right: 30 + 13 | bottom: 44

Piece 2: top: 68 | left: 43 | right: 24 | bottom: 10 + 22

Piece 3: top: 81 | left: 14 + 10 | right: 33 + 42 | bottom: 16 + 30

Piece 4: top: 30 + 14 | left: 27 + 62 | right: 20 + 21 | bottom: 19 + 30

Piece 5: top: 32 | left: 41 | right: 39 | bottom: 34

Piece 6: top: 46 | left: 19 + 20 | right: 33 | bottom: 23

Piece 7: top: 49 | left: 65 + 31 | right: 26 | bottom: 45 + 45

Piece 8: top: 14 + 20 | left: 16 + 10 | right: 38 | bottom: 96

Piece 9: top: 10 + 13 | left: 18 + 20 | right: 45 | bottom: 15 + 46

★ 49	42	43 + 24
28 + 75 40 + 49	89 57	37 + 20 28
30 + 43	26 + 40	28 + 40
73	66	68
64 47 + 10	57 42 + 20	62 19 + 43
13 + 40	48 + 10	41 + 10
53	58	51
37 + 62 31 + 30	61 29 + 40	69 46
95	70	27 + 56

★ **27 + 39** 16 ··· 13 + 60 50 + 41	**51** 73 ··· 17 + 50 36 + 60	**56** 67 ··· 20 + 27 26 + 50
91 73 + 23 ··· 79 23 + 70	**96** 29 + 50 ··· 12 + 50 19 + 70	**76** 62 ··· 73 21 + 50
93 39 ··· 86 38 + 46	**89** 16 + 70 ··· 80 + 14 23	**71** 94 ··· 85 84

★ (2 + 7) + 4	4 + (5 + 6)	1 + (8 + 2)
(3 + 4) + 5 ... 2 + (1 + 3)	(3 + 2) + 1 ... (5 + 2) + 3	5 + (2 + 3) ... 6 + (2 + 7)
(8 + 5) + 4	(7 + 0) + 5	4 + (1 + 6)
8 + (5 + 4)	(0 + 5) + 7	(1 + 4) + 6
8 + (6 + 1) ... 3 + (4 + 7)	7 + (3 + 4) ... 3 + (4 + 5)	5 + (3 + 4) ... 12 + (0 + 5)
(2 + 11) + 5	(1 + 0) + 8	9 + (3 + 7)
(5 + 2) + 11	0 + (8 + 1)	(3 + 9) + 7
(0 + 3) + 5 ... 9 + (6 + 10)	10 + (6 + 9) ... (7 + 2) + 4	(2 + 4) + 7 ... (14 + 3) + 7
(4 + 7) + 9	10 + (4 + 5)	8 + (6 + 4)

★ (8 + 5) + 4 (2 + 4) + 8 (3 + 4) + 5 (10 + 5) + 1	(7 + 0) + 5 5 + 7 1 + (2 + 6) 1 + (3 + 9)	4 + (1 + 6) 8 + 1 3 + (7 + 5) (4 + 5) + 2
15 + 1 (8 + 3) + 9 8 + (6 + 1) (4 + 7) + 9	12 + 1 8 + 7 (3 + 1) + 2 10 + (4 + 5)	2 + 9 4 + 2 10 + (2 + 3) 8 + (6 + 4)
11 + 9 6 + (4 + 0) (0 + 3) + 5 4 + (5 + 7)	9 + 10 5 + 3 (2 + 5) + 3 (6 + 1) + 3	10 + 8 3 + 7 (8 + 9) + 7 2 + (1 + 5)

Cut here ✂

★ 1 5 (2 + 4) + 5 (5 + 2) + 3	12 11 3 + (2 + 1) 4 + (3 + 2)	2 6 (8 + 19) + 1 (0 + 5) + 3
10 18 (9 + 1) + 5 (3 + 6) + 8	9 15 (7 + 3) + 4 6 + (5 + 2)	8 14 5 + (12 + 3) (8 + 1) + 9
17 7 10 + (7 + 3) 2 + (1 + 24)	13 20 (10 + 5) + 1 2 + (4 + 20)	18 16 (9 + 3) + 22 (18 + 3) + 6

This is a cut-apart matching puzzle. The cells contain the following numbers and equations:

★ **19** / **14** $(4+5)+6 = \underline{} + 5$ $(2+7)+4 = \underline{} + 7$	**17** / **10** $(9+7)+3 = 7 + \underline{}$ $4+(5+6) = \underline{} + 6$	**11** / **12** $3+(2+1) = 4 + \underline{}$ $1+(8+2) = 8 + \underline{}$
6 / **18** $(9+4)+1 = 9 + \underline{}$	**9** / **13** $(7+3)+6 = 3 + \underline{}$ $2+(6+3) = \underline{} + 3$	**3** / **7** $6+(1+10) = 10 + \underline{}$ $(7+3)+4 = 11 + \underline{}$ $2+(8+5) = \underline{} + 8$
5 / **15** $(8+3)+1 = 8 + \underline{}$ $(2+7)+4 = \underline{} + 4$	**8** / **4** $(10+9)+7 = \underline{} + 10$ $(4+2)+14 = 14 + \underline{}$	**16** $(10+5)+1 = \underline{} + 10$ $4+(6+8) = \underline{} + 4$

Cut here ✂

★ 1	17	10
14 0 + ___ = 0 2 + ___ = 5	0 4 + ___ = 10 5	6 20 2 + ___ = 13
3	5 + ___ = 10	11
16 7 + ___ = 20 1 + ___ = 18	13 12 5 + ___ = 19	4 + ___ = 16 2 8
17	14	0 + ___ = 8
8 4 + ___ = 8 15	4 6 + ___ = 13 12	7 19 9

★ 4	11	3
13 ... 15	$5 + \underline{\quad} = 20$ ... $1 + \underline{\quad} = 12$	11 ... 20
$6 + \underline{\quad} = 17$	4	$2 + \underline{\quad} = 10$
11	$1 + \underline{\quad} = 5$	8
9 ... $0 + \underline{\quad} = 16$	16 ... 17	$2 + \underline{\quad} = 19$ ... 7
8	$4 + \underline{\quad} = 9$	1
$7 + \underline{\quad} = 15$	5	$0 + \underline{\quad} = 1$
14 ... 10	$5 + \underline{\quad} = 15$ ... $3 + \underline{\quad} = 11$	8 ... 18
6	12	2

Cut here ✂

★

19 1 15

8 - ___ = 3

14 - ___ = 1

16 13 18

5

20 - ___ = 11   10   16 - ___ = 12

9   17 - ___ = 7   4

13 - ___ = 10

11 - ___ = 3

12 3 8 23

6   0 - ___ = 0   2

19 - ___ = 13   0   6 - ___ = 4

18 - ___ = 7

14 - ___ = 7

21 11 7 20

14   22   17

Cut here

★ 18 14 18 - ___ = 12	0 19 - ___ = 10 9 10	16 20 - ___ = 13 7 19 2
6 21 19 - ___ = 16 13	14 - ___ = 4 3 12 12 - ___ = 7	6 - ___ = 4 16 - ___ = 4 23 4
18 - ___ = 5 17 8 - ___ = 0 1	5 8 20	14 - ___ = 10 19 - ___ = 8 11 22 15

★ 8 9 2 + 3 7 + 8	10 5 19 + 1 11 + 5	3 20 2 3 + 4
15 5 9 + 5 14 11 + 9	16 1 10	7 0 + 1 6 12
20 0 3 1	6 + 4 1 + 2 12	2 + 10 11 + 8 19 4 7

A 3×3 grid of square puzzle pieces. Each piece has a value on each of its four sides (rotated to face outward).

	Column 1	Column 2	Column 3
Row 1	★ top: 19 · left: 9 · right: 2 + 2 · bottom: 9 + 6	top: 1 · left: 4 · right: 5 + 6 · bottom: 12 + 5	top: 15 · left: 11 · right: 2 · bottom: 5 + 4
Row 2	top: 15 · left: 5 · right: 2 · bottom: 0 + 0	top: 17 · left: 1 + 1 · right: 7 · bottom: 8 + 5	top: 9 · left: 2 + 5 · right: 6 · bottom: 3
Row 3	top: 0 · left: 0 · right: 4 + 2 · bottom: 1	top: 13 · left: 6 · right: 18 · bottom: 12	top: 3 + 0 · left: 10 + 8 · right: 4 · bottom: 7

★ / 2 (18, 20−0, 6)	17 (20, 3−0, 7)	13 (3, 23, 18−3)
18−12 (14, 14−3, 8−7)	15−8 (11, 9, 5)	15 (12−3, 19, 10−6)
1 (21, 14−2, 8)	7−2 (12, 19−19, 22)	4 (0, 10, 16)

Cut here ✂

Cut here ✂

★ 21 — 16 / 16 − 1 / 7	4 — 15 / 0 − 0 / 0 / 20 − 3	12 — 18 / 6
18 − 11 — 11 / 4 − 3 / 5	17 — 1 / 2 / 20 − 7	9 − 3 — 11 − 9 / 8 / 3
8 − 3 — 22 / 19 / 20	13 — 19 − 0 / 11 − 1 / 9	6 − 3 — 10 / 23 / 14

★ 12:30

6:30

5:00

10:30

11:00

4:30

7:00

8:30

2:30

9:00

6:00

10:00

12:00

3:00

1:30

2:00

7:30

11:30

4:00

1:00

3:30

8:00

5:30

9:30

Cut here

★

6 + 6	5 + 4	9 + 3
7 + 3 — 15 + 9	9 + 15 — 2 + 8	8 + 2 — 5 + 3
11 + 3	3 + 6	5 + 9
3 + 11	6 + 3	9 + 5
8 + 3 — 8 + 4	4 + 8 — 9 + 7	7 + 9 — 3 + 7
8 + 7	11 + 6	4 + 2
7 + 8	6 + 11	2 + 4
2 + 6 — 9 + 2	2 + 9 — 3 + 12	12 + 3 — 4 + 12
3 + 5	2 + 3	9 + 4

Row 1

	★ Top: 2 + 4 + 1	Top: 9 + 5 + 2	Top: 5 + 7 + 3
Left	6 + 11 + 4	1 + 0 + 2	2 + 5 + 3
Right	2 + 1 + 0	5 + 3 + 2	7 + 3 + 8
Bottom	6 + 8 + 2	3 + 4 + 1	4 + 6 + 1

Row 2

	Top: 2 + 6 + 8	Top: 1 + 3 + 4	Top: 4 + 1 + 6
Left	9 + 8 + 4	8 + 4 + 1	5 + 6 + 7
Right	4 + 1 + 8	7 + 6 + 5	8 + 5 + 3
Bottom	9 + 8 + 11	10 + 8 + 3	9 + 2 + 4

Row 3

	Top: 8 + 11 + 9	Top: 3 + 10 + 8	Top: 2 + 9 + 4
Left	7 + 5 + 8	6 + 7 + 12	9 + 5 + 15
Right	7 + 12 + 6	5 + 9 + 15	6 + 2 + 10
Bottom	3 + 12 + 7	15 + 5 + 4	2 + 5 + 13

★ **13** / 16 $5 + 8 = 8 + \_\_$ $\_\_ + 3 = 3 + 9$	**15** / 5 $3 + 7 = \_\_ + 3$ $8 + 0 = \_\_ + 8$	**12** / 7 $3 + 2 = \_\_ + 3$ $1 + \_\_ = 4 + 1$
9 / 11 $2 + 3 = \_\_ + 2$ $5 + 4 = 4 + \_\_$	**0** / 3 $7 + \_\_ = 2 + 7$ $\_\_ + 3 = 3 + 6$	**4** / 2 $7 + 6 = 6 + \_\_$ $5 + \_\_ = 1 + 5$
5 / 14 $9 + \_\_ = 8 + 9$ $10 + \_\_ = 4 + 10$	**6** / 8 $\_\_ + 5 = 5 + 10$ $\_\_ + 1 = 9 + 1$	**1** / 10 $4 + \_\_ = 6 + 4$ $5 + \_\_ = 8 + 5$

★ 16 + 35	64	39 + 52
50 + 27 67 + 7	74 50	42 + 8 41
76	54 + 6	75 + 4
72 + 4	60	79
37 96	89 + 7 81	79 + 2 43 + 72
49 + 9	86 + 8	82 + 6
58	94	88
24 + 53 68	61 + 7 55	49 + 6 69
47	82	64 + 23

★ 58	62 + 23	25
34 + 61 81 + 6 99	87 63 + 5 51	68 45 + 32 65 + 5
92 + 7	47 + 4	70
40 73 + 4 71	77 92 56	88 + 4 97 62
62 + 9	48 + 8	54 + 8
73 80 37 + 52	73 + 7 86 34	81 + 5 17 + 44 65

★

Put the puzzle together in the correct order.

ANSWER KEY – Page 1

Page 7

twenty-four	15	eighteen
17 1 one	ten 10	20
nine	8	2
9	eight	two
twenty-one eleven	11 four	sixteen
five	12	7
5	twelve	seven
14 three	3 six 6	22
twenty-three	19	thirteen

Page 7

Page 9

four	12	one
7 thirteen 13	23 twenty-three	9
21	seventeen	15
twenty-one	17	fifteen
two nineteen 19	twenty-two 22	six
fourteen	24	20
14	twenty-four	twenty
11 eighteen 18	sixteen 16	10
five	8	three

Page 9

Page 11

★ six	one	eleven
thirteen 30	thirty seventy 70	eight
sixty	40	ten
60	forty	10
two 100 one hundred	one hundred twenty 120	five
90	50	one hundred ten
ninety	fifty	110
seven 20	twenty eighty 80	three
twelve	four	nine

Page 11

Page 13

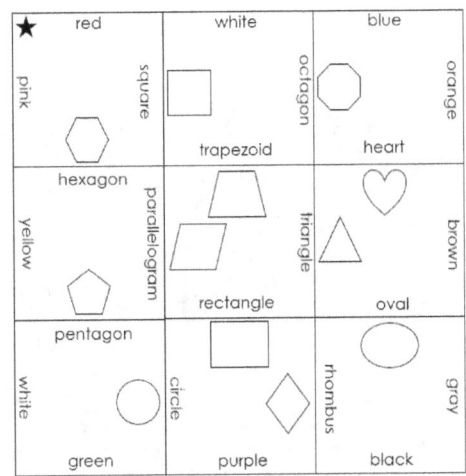

★ red	white	blue
pink square	octagon	orange
hexagon	trapezoid	heart
hexagon	parallelogram	triangle brown
yellow	rectangle	oval
pentagon		rhombus gray
white circle		
green	purple	black

Page 13

Page 15

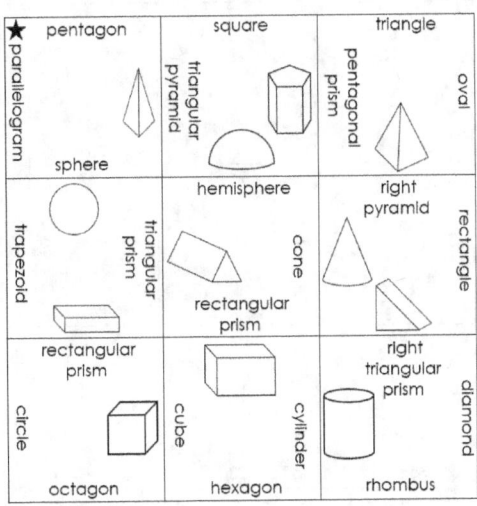

★ pentagon	square	triangle
parallelogram triangular pyramid	pentagonal prism	oval
sphere	hemisphere	right pyramid
trapezoid triangular prism	cone	rectangle
	rectangular prism	
rectangular prism		right triangular prism
circle cube	cylinder	diamond
octagon	hexagon	rhombus

Page 15

Page 17

★ 11	17	14
20 2 + 1 3	5 3 + 2	19
3 + 1	1	6
4	0 + 1	3 + 3
15 2 1 + 1	5 + 4 2	12
5 + 3	0 + 0	2 + 5
8	0	7
18 6 2 + 4	10 5 + 5	22
13	21	16

Page 17

ANSWER KEY – Page 2

Page 19

★ 20	14	18
11 2 + 8	10 2	0 + 2 12
2 + 2	3 + 4	7 + 2
4	7	9
16 6 + 3	9 8	4 + 4 15
1	6	3 + 7
1 + 0	4 + 2	10
21 1 + 2	3 4 + 1	5 19
13	17	0

Page 21

★ 17	22	15
12 4 − 2 2	0	1 − 1 19
5 − 2	9	1
3	10 − 1	3 − 2
16 10 − 0 0	5 − 1 4	11
9 − 2	7 − 1	5 − 0
7	6	5
21 10 − 9 1	10 − 0 10	13
18	14	20

Page 23

★ 22	15	13
16 6 − 4 2	4	10 − 6 19
8 − 2	6 − 5	7 − 4
6	1	3
12 9 − 6 3	5	8 − 3 14
0	7	8 − 6
7 − 7	10 − 3	2
20 9 − 0 9	10 − 2 8	21
17	18	11

Page 25

★ 21 + 52	68	81
58 30 + 13 43	24	14 + 10 33 + 42
44	10 + 22	16 + 30
30 + 14	32	46
27 + 62 20 + 21 41	39	19 + 20 33
19 + 30	34	23
49	14 + 20	10 + 13
65 + 31 16 + 10 26	38	18 + 20 45
45 + 45	96	15 + 46

Page 27

★ 49	42	43 + 24
28 + 75 40 + 49 89	57	37 + 20 28
30 + 43	26 + 40	28 + 40
73	66	68
64 47 + 10 57	42 + 20 62	19 + 43
13 + 40	48 + 10	41 + 10
53	58	51
37 + 62 31 + 30 61	29 + 40 69	46
95	70	27 + 56

Page 29

★ 27 + 39	51	56
16 13 + 60 73	17 + 50 67	20 + 27
50 + 41	36 + 60	26 + 50
91	96	76
73 + 23 29 + 50 79	12 + 50 62	73
23 + 70	19 + 70	21 + 50
93	89	71
39 16 + 70 86	80 + 14 94	85
38 + 46	23	84

ANSWER KEY – Page 3

Page 31

★ (2 + 7) + 4	4 + (5 + 6)	1 + (8 + 2)
(3 + 4) + 5 (3 + 2) + 1 2 + (1 + 3)	(5 + 2) + 3	6 + (2 + 7) 5 + (2 + 3)
(8 + 5) + 4	(7 + 0) + 5	4 + (1 + 6)
8 + (5 + 4) 3 + (4 + 7) 7 + (3 + 4)	(0 + 5) + 7 3 + (4 + 5)	(1 + 4) + 6 12 + (0 + 5) 5 + (3 + 4)
(2 + 11) + 5	(1 + 0) + 8	9 + (3 + 7)
(5 + 2) + 11 (0 + 3) + 5 9 + (6 + 10)	0 + (8 + 1) 10 + (6 + 9) (7 + 2) + 4	(3 + 9) + 7 (14 + 3) + 7 (2 + 4) + 7
(4 + 7) + 9	10 + (4 + 5)	8 + (6 + 4)

Page 33

★ (8 + 5) + 4	(7 + 0) + 5	4 + (1 + 6)
(2 + 4) + 8 (3 + 4) + 5 5 + 7	1 + (2 + 6) 8 + 1	3 + (7 + 5)
(10 + 5) + 1	1 + (3 + 9)	(4 + 5) + 2
15 + 1 (8 + 3) + 9 8 + (6 + 1)	12 + 1 8 + 7 (3 + 1) + 2	2 + 9 4 + 2 10 + (2 + 3)
(4 + 7) + 9	10 + (4 + 5)	8 + (6 + 4)
11 + 9 6 + (4 + 0) (0 + 3) + 5 5 + 3	9 + 10 (2 + 5) + 3 3 + 7	10 + 8 (8 + 9) + 7
4 + (5 + 7)	(6 + 1) + 3	2 + (1 + 5)

Page 35

★ 1	12	2
5 (2 + 4) + 5	11 3 + (2 + 1) 6	(8 + 19) + 1
(5 + 2) + 3	4 + (3 + 2)	(0 + 5) + 3
10	9	8
18 (9 + 1) + 5	15 (7 + 3) + 4 14	5 + (12 + 3)
(3 + 6) + 8	6 + (5 + 2)	(8 + 1) + 9
17	13	18
7 10 + (7 + 3) 20	(10 + 5) + 1 16	(9 + 3) + 22
2 + (1 + 24)	2 + (4 + 20)	(18 + 3) + 6

Page 37

★ 19	17	11
14 (4 + 5) + 6 = ___ + 5	10 (9 + 7) = 7 + ___	12 3 + (2 + 1) = 4 + ___
(2 + 7) + 4 = ___ + 7	4 + (5 + 6) = ___ + 6	1 + (8 + 2) = 8 + ___
6	9	3
18 (7 + 3) + 6 = 3 + ___	13 7	6 + (1 + 10) = 10 + ___ (7 + 3) + 4 = 11 + ___
(9 + 4) + 1 = 9 + ___	2 + (6 + 3) = ___ + 3	7
5	8	2 + (8 + 5) = ___ + 8
15 (8 + 3) + 1 = 8 + ___ 4	(10 + 9) + 7 = ___ + 10 16	(10 + 5) + 1 = ___ + 10
(2 + 7) + 4 = ___ + 4	(4 + 2) + 14 = 14 + ___	4 + (6 + 8) = ___ + 4

Page 39

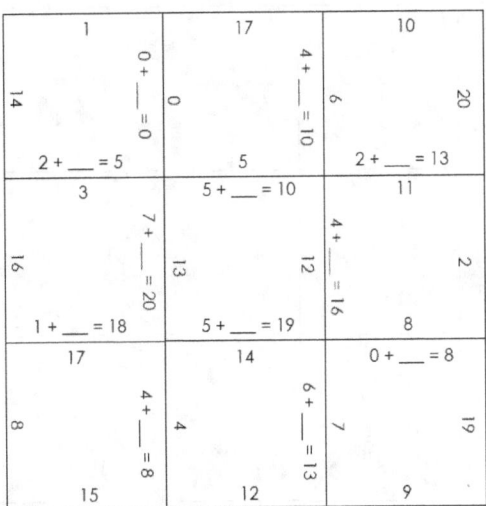

1	17	10
14 0 + ___ = 0	0 4 + ___ = 10 6	20 2 + ___ = 13
2 + ___ = 5	5	
3	5 + ___ = 10	11
16 7 + ___ = 20 13		4 + ___ = 16 12 2
1 + ___ = 18	5 + ___ = 19	8
17	14	0 + ___ = 8
8 4 + ___ = 8	4 6 + ___ = 13	7 19
15	12	9

Page 41

4	11	3
13 5 + ___ = 20 15	1 + ___ = 12 11 4	20 2 + ___ = 10
6 + ___ = 17		
11	1 + ___ = 5	8
9 0 + ___ = 16 16	17	2 + ___ = 19 7
8	4 + ___ = 9	1
7 + ___ = 15	5	0 + ___ = 1
14 5 + ___ = 15 10	3 + ___ = 11 8	18
6	12	2

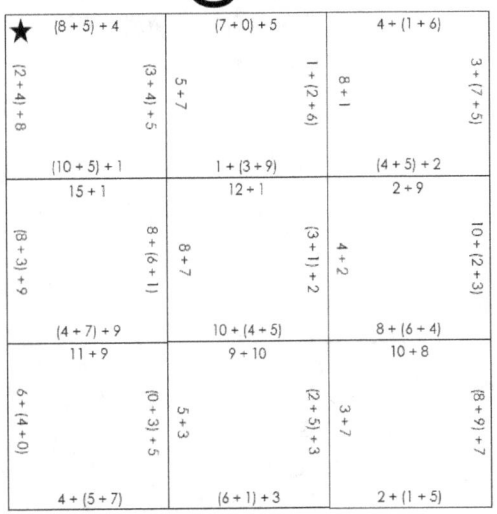

ANSWER KEY – Page 4

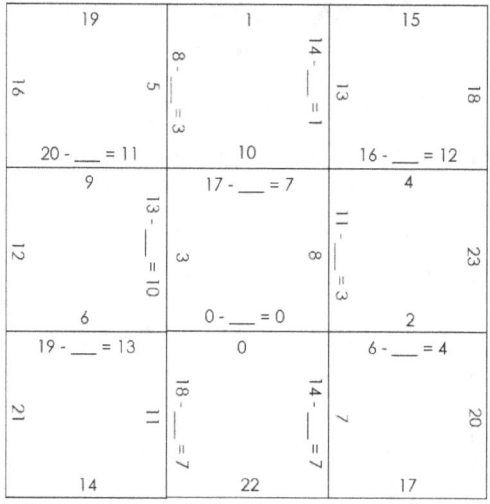

Page 43

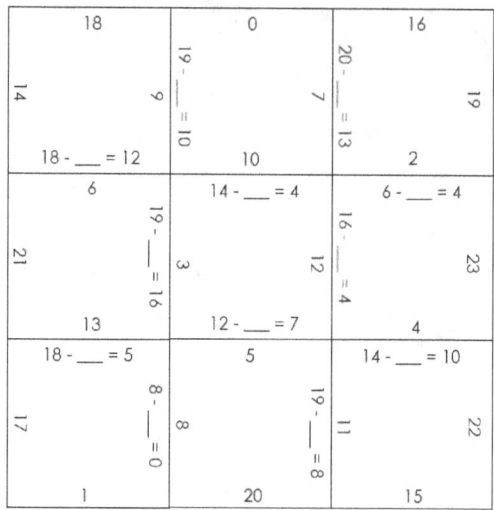

Page 45

Page 47

Page 49

Page 51

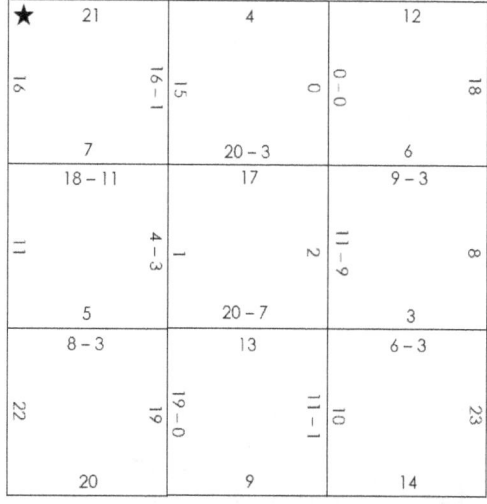

Page 53

ANSWER KEY – Page 5

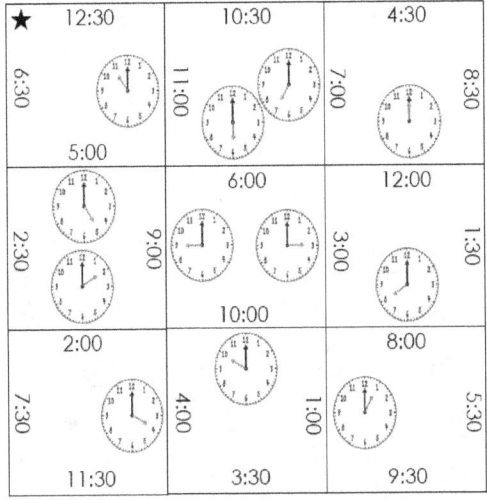

Page 55

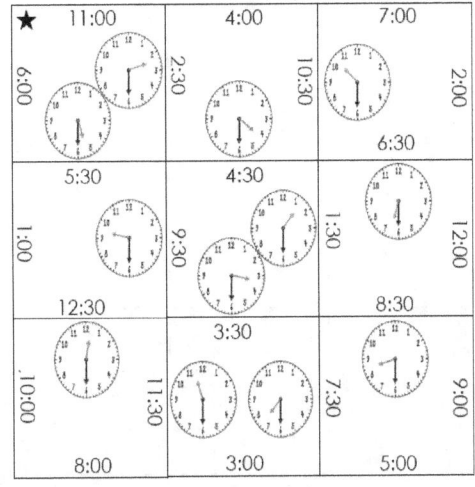

Page 57

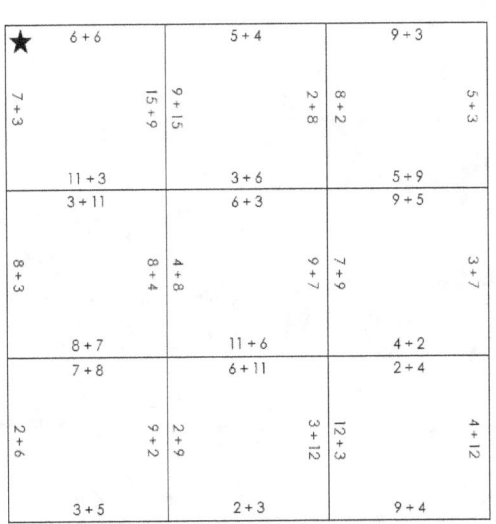

Page 59

★ 58 / 34 + 61 / 81 + 6 / 99	62 + 23 / 87 / 63 + 5 / 51	25 / 68 / 45 + 32 / 65 + 5
92 + 7 / 40 / 73 + 4 / 71	47 + 4 / 77 / 88 + 4 / 56	70 / 92 / 97 / 62
62 + 9 / 73 / 73 + 7 / 37 + 52	48 + 8 / 80 / 81 + 5 / 34	54 + 8 / 86 / 17 + 44 / 65

Page 67

www.HoJosTeachingAdeventures.com

Want FREE math puzzles you can use today?

SCAN ME